B. P. Pratten

Thousand Islands, Archipelago of the St. Lawrence River

B. P. Pratten

Thousand Islands, Archipelago of the St. Lawrence River

ISBN/EAN: 9783337240943

Printed in Europe, USA, Canada, Australia, Japan

Cover: Foto ©berggeist007 / pixelio.de

More available books at **www.hansebooks.com**

HINTS FOR PLEASURE SEEKERS.

THE

THOUSAND ISLANDS,

THE ARCHIPELAGO OF THE

ST. LAWRENCE RIVER.

BY ONE WHO HAS BEEN THERE.

WATERTOWN, N. Y.
DAILY TIMES PRINTING AND PUBLISHING HOUSE.
1889.

"And they were happy, and well content, sailing the way the river went."

THE
THOUSAND ISLES.

BY HON. CALEB LYON.

THE Thousand Isles, The Thousand Isles,
  Dimpled the wave around them smiles,
   Kissed by a thousand red-lipped flowers,
Gemmed by a thousand emerald bowers,
A thousand birds their praises wake,
By rocky glade and plumy brake,
A thousand cedars' fragrant shade
Falls where the Indians' children played,
And Fancy's dream my heart beguiles
While singing thee, The Thousand Isles.

The flag of France first o'er them hung,
The mass was said, the vespers sung;
The friars of Jesus hailed the strands
As Blessed Virgin Mary's lands;
The red men mutely heard, surprised,
Their heathen names all christianized,
Next floated a banner with cross and crown;
'Twas Freedom's eagle plucked it down,
Retaining its pure and crimson dyes
With stars of their own, their native skies.

There St. Lawrence gentlest flows,
There the south-wind softest blows,
There the lilies whitest bloom,
There the birch hath leafiest bloom,
There the red deer feed in spring,
There doth glitter wood duck's wing,
There leap the muscalonge at morn,
There the loon's night song is borne,
There is the fisherman's paradise,
With trolling skiff at red sunrise.

The Thousand Isles, The Thousand Isles,
Their charm from every care beguiles,
Titian alone hath power to paint
The triumph of their patron saint,
Whose waves return on Memory's tide;
LaSalle and Piquet, side by side
Proud Frontenac and bold Champlain,
There act their wanderings o'er again;
And while their golden sunlight smiles,
Pilgrims shall greet thee, Thousand Isles.

HON. R. A. LIVINGSTON'S ISLAND

A SUMMER PARADISE.

THE THOUSAND ISLANDS.

NATURE nowhere presents more alluring charms than in that labyrinth of land and water known as The Thousand Islands of the St. Lawrence. In the old Indian days this beautiful extent of the river was called Manatoana, or Garden of the Great Spirit, "and well might the islands, when covered with thick forests, the deer swimming from wooded isle to wooded isle, and each little lily-padded bay nestling in among the hills and bluffs of the islands, and teeming with waterfowl, seem to the Indian, in his half poetic mood, like some beautiful region dedicated to his Supreme Deity."

HISTORICAL.

This region has a history full of romantic interest, as any one can imagine when he remembers that it has four times been the border land between contending nations. First, between the two great Indian races, the Algonquins and the Iroquois ; next, between the French and the English, and twice between the English and Americans. But our space is too limited for more than a few items.

The St. Lawrence was discovered by Jacques Cartier, a good Catholic, on St. Lawrence's day, in 1535 : hence the name. Fort Carleton, the ruins of which are seen upon the upper end of Carleton Island, just below Cape Vincent, was built in the beginning of the Revolutionary war by the British commander, General Carleton. It was the principal military station above Montreal, and remained in the possession of the British until the beginning of the war of 1812. The boundary line between Canada and the United States, which runs

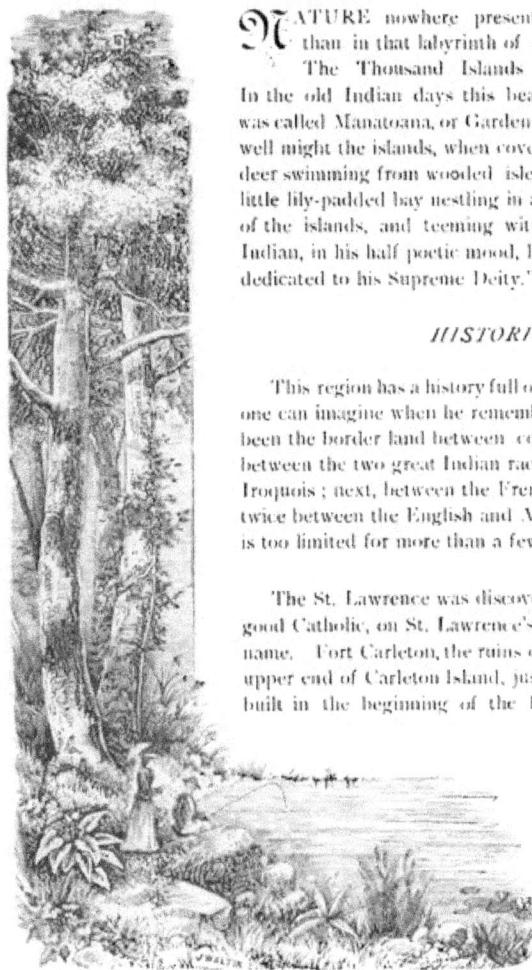

through these islands, was not definitely settled until 1822. The first steamboat on the St. Lawrence was the Oneida, in 1817. It caused great excitement along the shores.

The Patriot war, a Canadian outbreak, which led to exciting military adventures on the St. Lawrence, occured in 1837 to 1839. During this war, the British steamer, Sir Robert Peel, was fired and burned on the south side of Wells Island, on the night of May 20th, 1838, and the "Battle of the Windmill" occured near Prescott in November of the same year, a memorable battle to the elder Crossmon, who was taken prisoner during the engagement, tried and sentenced to be shot. Owing to his extreme youth, a respite was obtained, and he was afterwards ransomed, thus barely escaping with his life.

GEOLOGICAL.

THE WINDMILL.

The geological formation of the Thousand Islands is mostly gneiss rock of the Laurentian period. The rock is composed largely of a reddish feldspar, with mixtures of quartz and hornblende, and a little magnetic iron ore. There are also occasionally thin veins of trap and greenstone, and in places a variety of crystalline mineral forms. Potsdam sandstone occurs among the islands in thick masses, rising sometimes into high cliffs. Before reaching Brockville from above, and for a long distance below, a calciferous sandstone and the older limestones continue the only rock, and in these are found the organic remains of lower forms of animal and vegetable life.

DESCRIPTIONS OF THE ISLANDS BY DIFFERENT WRITERS.

The first mention of the Thousand Islands was by Samuel Champlain, who visited Lake Ontario and the Upper St. Lawrence in 1615. In his meagre descriptions he mentions some beautiful and very large islands at the beginning of the St. Lawrence. It is supposed that some French explorers, who went up the river about 1650, gave the region its present name, "*Milles Isles*," or Thousand Islands. In the papers relating to DeCourcelle's and DeTracy's expeditions against the Indians, in 1666, the islands are complained of as "obstructing navigation and mystifying the most experienced Iroquois pilots."

In the year 1820, a Capt. Pouchot described the region somewhat minutely in his journal, which was afterwards published in Switzerland, and there have been frequent allusions to and descriptions of it written and published from that time to this. The picturesque scenery of this spot also seems to have made a lasting impression upon French artists, as one of the finest paintings which greet the eye of an American on entering the Picture Gallery at Versailles presents a view of these attractive wilds.

IN ROMANCE AND SONG.

We find them occasionally in poetry and fiction. "The Canadian Boat Song," by the great Irish poet, Thomas Moore, commencing :

> "Faintly as tolls the evening chime,
> Our voices keep time and our oars keep time,"

was written in 1804, it is said, on Hart's Island, opposite The Crossmon. During their passage down the river, James Fennimore Cooper and Washington Irving visited the Thousand Islands, and were fascinated by them. Cooper makes them the scenes of some of the most interesting incidents of "The Pathfinder," from which we copy the following description :

"By sunset again the cutter was up with the first of the islands that lie in the outlet of the lake, and ere it was dark she was running through the narrow channels on her way to the long-sought station. At 9 o'clock, however, Cap, insisted that they should anchor, as the maze of islands became so complicated and obscure that he feared, at every opening, the party would find themselves under the guns of a French fort. * * * The islands were so numerous and small as to baffle calculation, though occasionally one of a larger size than common was passed. Jasper had quitted what might have been termed the main channel, and was wending his way, with a good stiff breeze and a favorable current, through passes that were sometimes so narrow that there appeared to be barely room sufficient for the Scud's spars to clear the trees ; at other moments he shot across little bays, and buried the cutter again amid rocks, forest and bushes. The water was so transparent that there was no occasion for the lead, and being of very equal depth little risk was actually run."

Farther on he describes the island where "The Pathfinder" and his party secreted themselves, which is so good of many others that we insert it here :

"Lying in the midst of twenty others, it was not an easy matter to find it, since boats might pass quite near, and, by the glimpses caught through the openings, this particular island would be taken for a part of some other. Indeed, the channels between the islands that lay around the one we have been describing were so narrow that it was difficult to say which portions of the land were connected or which separated, even as one stood in their centre, with the express desire of ascertaining the truth. The little bay, in particular, that Jasper used as a harbor, was so embowered with bushes and shut in with islands that, the sails of the cutter being lowered, her own people, on one occasion, had searched for hours before they could find the Scud, on their return from a short excursion among the adjacent channels in quest of fish."

FROM THE CENTURY.

"Now, however, the inexorably rotating kaleidoscope of time has shaken away the savage scenes of old, never to be repeated, and new ones appear to the eye of the present. No longer in Alexandria Bay —fortunately still beautiful—does nature reign in silent majesty, for the constant flutter and bustle of the life and gayety of a summer resort have superseded her. But although Alexandria Bay is in the continual tumult of life, for some fortunate and almost unaccountable reason, the Thousand Islands are not in the least tinctured with the *blase* air of an ordinary watering-place, nor are they likely to become so. There are hundreds, thousands of places, rugged and solitary, among which a boat can glide while its occupant lies gloriously indolent, doing nothing but reveling in the realization of life; little bays, almost land-locked, where the resinous odors of hemlock and pine fill the nostrils, and the whispers of nature's unseen life seem but to make the solitude more perceptible. Sometimes the vociferous cawing of crows sounds through the hollow woods, or a solitary eagle lifts from his perch on the top of a stark and dead pine, and sails majestically across the blue arch of the sky. Such scenes occur on a beautiful sheet of water called Lake of the

SAFE POINT.

Isle, lying placidly and balmily in the lap of the piny hills of Wells Island, reflecting their rugged crests in its glassy surface, dotted here and there by tiny islands. In the stillest bays are spots that seem to lie in a Rip Van Winkle sleep, where one would scarcely be surprised to see an Indian canoe shoot from beneath the hemlocks of the shore into the open, freighted with a Natty Bumpo or a Chingachgook, breaking the placid surface of the water into slowly widening ripples. In such a spot, one evening after a day spent in sketching, when paddling our boat about in an indolent, aimless way, looking down through the crystal clearness of the water to the jangle of weeds below, now frightening a pickerel from his haunt or starting a brood of wood duck from among the rushes and arrow-heads, we found ourselves belated. As the sun set in a blaze of crimson and gold, two boatmen moving homeward passed darkly along the glassy surface, that caught the blazing light of the sky, and across the water came, in measured rythm with the dip of their oars, the tune of a quaint, old, half-melancholy Methodist hymn that they sang. We listened as the

FIDDLER'S ELBOW.

song trailed after them, until they turned into an inlet behind the dusky woods and were lost to view. From such romantic and secluded recesses one can watch the bustle and hurry of life as serenely as though one were the inhabitant of another planet."

IN RECENT LITERATURE.

During the past few years, wherein the Thousand Islands have suddenly become one of the leading resorts for summer recreation, they have been prominent in the current literature and pictorial illustrations of the country. Newspapers and magazines have made them the subject of many long and interesting articles; reporters, essayists, romancers, poets and humorists have seemed to vie with each other in calling the attention of the public to this place of enchantment ; and the consequence is, that a vast and annually swelling tide of humanity flows that way, and many linger there from early June until late October.

2

Fair St. Lawrence! What poet has sung of its grace
As it sleeps in the sun, with its smile-dimpled face
Beaming up to the sky that it mirrors? What brush
Has e'er pictured the charm of the marvelous hush
Of its silence, or caught the warm glow of its tints
As the afternoon wanes, and the even-star glints
In its beautiful depths?   And what pen shall betray

The sweet secrets that hide from man's vision away
In its solitudes wild?   'Tis the river of dreams;
You may float in your boat on the bloom-bordered streams,
Where its islands like emeralds matchless are set,
And forget that you live, and as quickly forget
That they die in that world you have left; for the calm
Of content is within you, the blessing of balm
Is upon you forever.—ANON.

CHARACTERISTICS.

THE ISLANDS AND CHANNELS.

There are nearly two thousand of these St. Lawrence islands, and perhaps one thousand within six miles of Alexandria Bay, this being the central part of by far the most beautiful and wonderful section of the river. They are nearly all small, usually varying in size from a few square yards of surface to several acres. Many of them are separated only by narrow channels, which are generally deep, but sometimes shallow. Quiet and inviting little bays are found here and there. All the islands are thickly studded with trees of rich foliage, but generally of moderate or stunted growth, many of which stand close to the water's edge, and afford cooling shade to passing boatmen. In the bays and by the sides of the islands is excellent fishing, bass and pickerel being the principal fish, but the famous muscalonge is sufficiently numerous to warrant the fisherman in expecting an electric bite from him at any moment, which will put his strength and skill to their utmost test.

WELLS ISLAND.

Special mention should here be made of the largest of the islands, the lower end of which is just below the village of Alexandria Bay. It is eight miles long, and from a few feet to four miles wide. Portions of it have been cultivated as farms for the last half a century. Other parts are charmingly wooded, and some of its rock features are exceedingly picturesque. The lower portion is separated into two parts by the " Lake of the Islands," which is

connected with the river on the American and Canadian sides by two narrow channels. This quiet lake, three or four miles long, is fringed with rich foliage and occasional bold rocks, and is a favorite fishing and hunting resort.

AS A SUMMER RESORT

OLD TIMES.

Not until 1872 was the attention of the general public turned to the Thousand Islands as a " watering place," or a resort for pleasure seekers and invalids, although some discerning ones had been in the habit of spending a few summer days or weeks there for more than a quarter of a century previous. There Gov. Seward shook hands across the party chasm with Silas Wright, and caught bass and muscalonge with him from the same boat, exchanging practical quotations and cheerful jokes instead of political opinions and arguments. There Rev. Dr. Geo. Bethune dropped theology, and Gen. Dick Taylor forgot military tactics, and floated sociably together down among the islands. There the wily Martin Van Buren and his witty son John, Frank Blair and other politicians of the old school found respite from the affairs of the State and partisan squabbles, and were soothed and softened by the influences of nature. And when these intellectual giants returned from their fishing expeditions they found rare good cheer and comfort in the unpretentious old Crossmon house at the Bay, where the elder Crossmon was then known as the prince of country landlords, and in such goodly company learned thoroughly the fine art of managing and entertaining guests.

INLET TO THE BAY.

In the summer of 1872, two or three things occurred opportunely to draw immediate attention to the river attractions. George W. Pullman, the palace-car king, had become enamored with the place, purchased a beautiful island nearly opposite the Bay, and erected thereon suitable buildings for a luxurious summer residence. By his invitation, in 1872, General Grant and family and a party of friends went to Pullman's Island, as his guests, and remained eight days. The same season a large party of New York and Southern editors made an excursion to the islands, and dined *al fresco* on the same island, the viands being furnished from the cuisine of the Crossmon House. These two events brought the islands to the notice of the people in all parts of the country.

So, when the big new hotels were opened in the summer of 1873, the people at once began to hasten to them, and since then they have continued to come every year in large numbers.

Among the distinguished visitors to the islands, a year ago last summer, were President Cleveland and party. Although their stay was of short duration, they were very much taken up with the beautiful scenery and the numerous islands.

CATCHING MUSCALONGE.

DISTINGUISHED GUESTS.

The Crossmon has been particularly honored of late by being the chosen stopping place of President Arthur, Gen. Sheridan, Cardinal McClosky, Herbert Spencer, Charles Dudley Warner, the artist Reinhart, Will Carlton and Marietta Holley.

FISHING PICNICS.

Several of these enjoyable affairs come off every pleasant day. A party of from 10 to 25 ladies and gentlemen set off in a steam yacht for some distant fishing-ground, taking liberal supplies from the hotel, and about half as many oarsmen as excursionists. Each oarsmen takes his own skiff and fishing tackle. The boats being towed in single file behind the yacht present the appearance of some strange marine animal, with a very long tail. An island is selected as the base of operations, and here the yacht is moored to the shore, and the party separate, each skiff, with its two or three occupants, taking a different direction, with the understanding to meet again at that island for dinner. At the appointed hour the boats return, and the oarsmen, nearly all of whom are good cooks, set at work preparing dinner. A fireplace is quickly improvised out of rocks, and the savory odors of a hot dinner soon mingle with the piny odors of the woods. The yacht carries boards for tables, and the island supplies rocks to support them. The afternoon is spent in rambles on the adjacent islands, or in story-telling under some big tree, while two or three drowsy gentlemen go off to sleep under the influence of the fresh air and a hearty dinner. Frequenters of the islands often bring hammocks with them for these occasions.

RIVER SPORTS.

Boating, fishing, hunting, cruising among the islands in row-boats or steam yachts, visiting many points of historical or traditionary interest, picnicking in large or small parties, open-air feasting, and lounging under the trees by the water's edge are terms which sum up the principal sports of the river. There are many small boats at the Bay, and many good oarsmen stand ready to serve, at a moderate price, those who want their services. These oarsmen are a convenience, but not a necessity, to the enjoyments among the islands. They know all the good fishing grounds, can give all needed instructions in the art of catching, will furnish the requisite fishing tackle, and cook the fish in dainty and appetizing style when caught.

DESCENDING THE RAPIDS.

Black bass and pickerel, large and gamy, abound in these waters. Many muscalonge are also caught every season, and the lady or gentleman who hooks and secures one or more of these largest and best of all fresh water fish becomes the heroine or hero of the day on returning to the Bay.

Ladies are often the lucky ones, and sometimes pull in a muscalonge of enormous size, courageously refusing the while all masculine assistance.

Occasionally a muscalonge weighs as high as forty pounds, a pickerel as high as twenty pounds, and a bass as high as six or seven pounds. The muscalonge

ALEXANDRIA BAY FROM BLUFF ISLAND

are mostly taken between the middle of May and the last of July, the bass bite best between the middle of June and September, while the pickerel are caught early and late in the season, and all the season.

Trolling is the usual and the most exciting method of fishing among the islands, though much pleasant still fishing is also done.

Late fall and early spring, as all hunters know, are the times for shooting duck, when they flock to the bays and coves of this section of the river by thousands. It is not unusual for a fishing party to return to the Bay at night with a hundred or more fine fish, nor for the hunter to come in with fifteen or twenty broad-billed trophies of his marksmanship.

STEAMERS AND YACHTS.

The large line steamers are seen plying up and down the river at frequent intervals. All of them touch at the Bay, and many others, nearly as large, are devoted to excursions. A new company has been formed to meet the demands and necessities of the increasing travel along the river and Lake Ontario by putting on a line of floating palaces, similar to those on the Hudson, and costing from $80,000 to $100,000 each.

An important feature of life at the Bay and among the Thousand Islands is the great and increasing number of steam yachts, large and small, which glide to and fro over the water and in and out among the island channels during the pleasure season. Some of these are models of architectural beauty, such as can be seen almost nowhere else. . In going considerable distances on the river, these yachts afford a swift and delightful conveyance for small parties, as the larger steamers do for larger parties, and for distant places and picnics, or extensive views of the river scenery.

THE WANDERER.

The best way of gaining a comprehensive idea of the magnitude and wildness of this archipelago is by taking a trip on the new Island Wanderer. This fast steamer makes two trips daily, of forty miles each, taking in on its way some of the most intricate channels among the islands.

3

A SCENE OF ENCHANTMENT.

The summer night scenes at the Bay are weirdly enchanting, and European travelers say they remind them of the night scenes at Venice, and are quite as beautiful. The illuminations extend far up and down the river, on gliding yachts and steamers, on the islands, along the grounds and in the windows and towers of the great hotels, and added to these are the lights of the village and nightly displays of Chinese lanterns, Roman candles, rockets and other fireworks. This superb kaleidoscope of river fires must be witnessed to be appreciated.

SOME NAMES EXPLAINED.

The historically famous Devil's Oven is an island so named from a water cave into which a boat can be rowed from the river. This cave was the hiding place for many months of the famous "Bill Johnson" during the Patriot War of 1837–39. Aided and sustained by his daughter Kate, he finally escaped.

Goose Bay is a well-known fishing and hunting ground three miles from the village. Eel Bay is another at the head of Wells Island. Halstead's Bay is another on the Canadian side.

Fiddler's Elbow is a thick and favorite cluster of islands in the Canadian waters.

THE RIFF.

The Riff is the long, narrow inlet to the Lake of the Islands. It is over a mile long, and so narrow that a child can throw a stone across it at any point, and yet is navigable for small yachts.

THE COTTAGES.

The river cottages are numerous, and every year important additions are made to them. It is noticeable that as time passes the new ones constructed are more and more costly.

PARKS.

Round Island Park occupies a large island nine miles above Alexandria Bay. It belongs to a Baptist Association, which was organized in the summer of 1879.

The Thousand Island Park of the Methodists is on the upper end of Wells Island, two miles below Round Island. It was started in 1873, and to its natural beauties have been added delightful drives and walks, a village of cottages, bath-houses, and buildings for religious purposes and the accommodation of visitors. Here are held Sunday school, temperance and educational conventions every season.

Westminster Park is at the foot of Wells Island, about a mile and a half from the Bay. It was purchased in 1874 by a Presbyterian stock company, and has been rapidly improved, having now several miles of drives and some fine buildings.

Edgewood Park, on the mainland opposite the Bay, is owned by the Edgewood Park Association of Cleveland, Ohio. This Association is composed of gentlemen of means, who, with their families, wish to spend a few weeks at the islands each year. The Park comprises thirty acres of wooded land. A club house and some cottages have already been erected for the comfort and convenience of the club, and a large sum of money expended in beautifying the Park.

These Parks are connected with each other and the Bay many times daily by steamers, which afford delightful little trips.

Although the Thousand Islands are now dotted with cottages, and thronged here and there with people, their original wild beauty and enticements remain — the trees and rocks ; the majestic flow of crystal pure waters ; the yet purer air, with its splendid tonic and healing properties ; the ever-varying views ; the opportunities for boating, fishing, hunting, bathing, etc. all are here, and man has added to them yet more.

DRIED GRASSES FROM THE ISLANDS.

Extensive improvements, are constantly being added to the parks and islands. Several beautiful cottages have recently been built and still more are in contemplation. Owners of property on the river take considerable pride in beautifying and fitting up their cottages, and making them attractive and comfortable.

PICNIC DINNER ON AN ISLAND

THE CAPES.

Many small capes, which scallop the main shores of the river, afford beautiful building sites, and some of them are adorned by handsome cottages. The demand for these capes has increased of late, and it is probable that before many years the shores for a long distance each way from the Bay, as well as the islands, will be thickly studded with cottages, owned by health and pleasure seekers from abroad.

Perhaps the most desirable point on the river was purchased by Dr. J. G. Holland, the celebrated author, and late editor of The Century. It is at the mouth of the lovely little bay overlooked by the Crossmon, and only a few rods from it across the water. Dr. Holland has expended many thousands of dollars in erecting here a

DR. J. G. HOLLAND'S LATE RESIDENCE, "BONNIE CASTLE."

luxurious cottage and improving the grounds. The point is named "Bonnie Castle," from one of Dr. Holland's novels. The family are in the habit of spending three to four months of the year on this island, and here Dr. Holland did much of his literary work.

Below "Bonnie Castle" are the Ledges, owned by Mrs. Sara E. K. Hudson, of New York, which have recently been laid out in flower beds and lawns. The Ledges are on the main land, and the recent improvements add considerable to the beauty of the locality.

ISLAND ROYAL.

PARTICULAR ISLANDS.

In 1823 all the islands on the American side, between Ogdensburg on the St. Lawrence and Grindstone Island in Lake Ontario, were granted to Elisha Camp, of Sackets Harbor, and all titles within these limits must be traced to this proprietor.

Island Royal, owned by Mr. Royal E. Deane of New York, is situated opposite Point Vivian, two miles from Alexandria Bay, and quite near Wells Island. The veranda of the cottage is twenty feet above the water, and from this elevation a view unsurpassed upon the channel may be enjoyed. Many of the river captains pass within hailing distance of this beautiful spot. Mr. Deane and family for many years have been summer residents upon the river.

Just above the village, in the American channel, is Warner Island, owned by H. H. Warner, of Rochester, who is famous for the magnitude, boldness and success of his business operations. The line steamers pass within a few feet of his cottage. Mr. Warner and family are in the habit of remaining here two or three months of the year, and their gracious hospitalities have won them hosts of friends among the frequenters of the Thousand Islands.

Across the channel, on Pullman's Island, already referred to, stands a magnificent castle built of rough, unhewn stone, designed after a castle on the Rhine, and very appropriately called " Castle Rest." It stands on the site of the earlier cottage, where President Grant was entertained.

The buildings and grounds are lighted by electricity, and from the lofty tower a fine view of the surrounding islands can be obtained. This beautiful island, with its summer palace, was presented by Mr. Pullman to his mother on her eightieth birthday.

Near by is Nobby, which, owing to its position and natural formation, is one of the most desirable among the islands. The owner, H. R. Heath of New York, has devoted much time and capital in improvements, both on Nobby and the famous Devil's Oven, which is also in his possession.

Point "Marguerite"

A short distance down the river, and opposite Nobby, is Friendly Island, owned by Mr. E. W. Dewey, of New York. The natural beauty and location of the island, combined with the elegance and taste of the house and surroundings, make this one of the most charming and attractive of the summer homes.

Rye Island has recently been purchased by Nathaniel W. Hunt, of Brooklyn, and re-christened St. Elmo. The cottage is a prominent one, and is the design of the architect who has built most of the finest cottages on the river.

Opposite The Crossmon is Isle Imperial, in some respects the most remarkable of the inhabited islands. When purchased, in 1882, it consisted of a mere rock, a few square yards in extent, but by piering and filling in the owner has increased the size to half an acre, and thus obtained an island in one of the most desirable locations on the river. It is now owned by Charles I. Singer, of Chicago, Ill. Near by is Hart's Island, already mentioned.

Plantagenet Island was purchased by Judge Charles Donohue, of New York, and re-christened " St. John." He has built a handsome cottage upon it, and is constantly making improvements.

The details might be indefinitely extended.

A short distance down the river from the Bay is a triplet of charming little islands They are : Little Lehigh and Sport, owned by E. P. Wilbur, Bethlehem, Pa., and Idle-wild, owned by Mrs. Eggleston, New York. The first two are connected by a hand-some wrought iron bridge.

Sport Island is nicely terraced, and a private gas house furnishes the means of illuminating it at night with two hundred lights.

GRAND HARBOUR ISLAND

ALEXANDRIA BAY.

This village is the central point of interest, from its nearness to the most picturesque part of the islands. It has a population of about seven hundred, and is prettily situated on a point of land between two river bays, making it almost water bound. The fishing in this vicinity is better than elsewhere, owing to the greater number of islands, which cause quiet shallows, where fish delight to congregate. Here, too, is the

CHURCH OF THE THOUSAND ISLANDS,

built in 1851, through the instrumentality of Rev. Dr. George W. Bethune, of the Reformed Dutch church, who was a regular visitor at the Bay for many successive years, commencing as early as 1845. The church building, which is a chaste stone structure, with truncated tower, stands on a knoll in the edge of the village.

METHODIST CHURCH.

There is also a pretty little church, recently completed by the Methodists, at a cost of about $6,000, finished inside in black walnut and ash, and nicely carpeted. It has a capacity for seating about 300 persons.

THE NEW EPISCOPAL CHURCH.

Through the efforts of Bishop Huntington and others, an Episcopal chapel is being erected at the Bay, and will be completed at the opening of the present season.

LIBRARY.

A fine library has been established at the Bay for the use of visitors, under the auspices of the Y. M. C. A. It has about one thousand volumes, a large portion of which were generously donated by its founder, Dr. Holland. These will be increased from year to year.

"WILD FLOWERS OF THE ISLANDS."

IMPROVEMENTS.

For some time past noticeable improvements have taken place at the Bay, and although it will be some years before it will assume the proportions of a large town or city, it is slowly but surely growing. It is widely known as a summer resort. Recently new houses have been erected, old ones painted and otherwise improved, new walks built, etc. With its large hotels, its fine residences, its location, (the best on the river,) its healthy locality, its thousands of visitors, Alexandria Bay is the principal place to live during the hot summer months.

DEVIL'S OVEN

NEWPORT ISLAND.

THE CHILDREN'S BURRO BRIGADE AND TENNIS COURT. (STREET SIDE OF THE CROSSMON.)

THE CROSSMON

We now come to that which provides sweet and invigorating rest after the varied river sports, country drives and sociabilities. We mean THE CROSSMON. The old hotel, under the same management as the new, has been referred to. It had been the stopping place for visitors to the islands for more than a quarter of a century, and acquired during that time a reputation of which any hotel with similar facilities might be proud. The new, many-towered Crossmon consists of a five-story building, covering exactly the site of the old hotel of pleasant memories. It is a picturesque structure, surrounded by wide verandas and traversed by spacious halls.

THE SITUATION AND OUTLOOK

It is most charmingly situated, close to the river on the north, and the little gem of a bay from which the village takes its name on the east, thus having *two water sides*. Its windows, verandas and towers afford extensive views of the river and islands in three directions. Most of the prominent islands and cottages may be seen from it, together with miles and miles of the sweeping, bounding, gleaming river. The hotel has in reality two fronts, (with their entrances,) the one being toward the river, where boat passengers enter, and the other on the main village street, where carriages are the mode of conveyance.

PRINCIPAL ADVANTAGES OF THE CROSSMON.

The office, wine room, billiard room and barber shop, being on the street front of the hotel, are entirely removed from the water front, where the verandas are, and where the guests like to assemble for games and promenading. An elevator runs from the basement to the top of the building, and the broad stairways, in both main building and wing, afford quick means of egress in case of fire.

THE CROSSMON IN 1878

The hotel is lighted throughout with gas, and supplied with pure river water, which is forced by a steam engine into an enormous *copper tank* on the roof, and conveyed from there to the various floors by means of *galvanized iron pipes*, thus doing away with all danger of *lead poisoning* and other impurities. On every floor are water-closets and bath-rooms, with hot and cold water. Electrical bells and speaking tubes connect the office with every part of the building. It will accommodate three hundred guests, and is adapted to satisfy those who are accustomed to luxurious homes.

THE CROSSMON IN 1863.

The table is supplied with all the delicacies of the season, prepared by accomplished cooks; and the best brands of foreign wines, beers and liquors await the orders of all who desire them.

Morning concerts are given by a fine orchestra, and the amusements of the day are varied in the evening by music, dancing and games in the parlor, and thus the round of enjoyment may be continued from early morning until late bed-time. There are over five hundred feet of verandas, and guests may promenade the entire distance, and through the long halls, without obstruction.

Appetizing lunches are neatly put up free of charge at the hotel for picnic and fishing parties, and, after a ride on the river, are often enjoyed in the open air, under the trees, even better than the most sumptuous dinners in the dining-rooms. Boats, oarmen and fishing tackle can be engaged for parties wishing them by applying at the hotel office.

Considerable attention has been given to provide for the entertainment and amusement of the children of our guests. Four Rocky Mountain burros have been imported

by us from New Mexico for their use.   These little animals are very gentle, and are
trained for riding and driving.   For the smaller children, a goat trained for driving is
provided.

The grounds of the hotel, over four acres in extent, have been grassed and nicely
graded, and are beautified in places by beds of flowers.

On the east, towards the bay, is an extensive lawn, reaching to the water's edge.
On this side is the principal landing place for yachts and smaller boats.   On the north

THE CROSSMON IN 1874.

is a rocky incline, spotted with grass and flowers.   The grounds, as well as the buildings, are brilliantly illuminated at night, colored lights shining in all the towers, which
have a peculiarly beautiful effect, as seen from the river.

Notwithstanding the extensive accommodations, The Crossmon is crowded much of
the time during the warm season, and it is therefore a good plan for parties wishing
rooms to engage them in advance by letter, or through the agency of friends.

Address,

C. CROSSMON & SON,
THE CROSSMON,
ALEXANDRIA BAY, N. Y.

NAMES OF ISLANDS AND POINTS

The following are the names of the inhabited islands and points, beginning in order at Clayton, and extending below Alexandria Bay:

CEMENT—(Point Head, Grindstone Island,) eighty acres, owned by.. ...W. F. Ford, Latargeville, N. Y.
GOOSE ISLAND—two acres, owned by .... E. S. Hicks, Brooklyn, N. Y.
HEN ISLAND—one-quarter acre, owned by, ...W. F. Morgan, New York
DAVITTS' ISLAND—one-quarter acre, owned by, ...H. G. Davitts, New York
CORAL ISLE—two acres, owned by ....C. Wolfe, New York
FAIR VIEW POINT—one acre, owned by... .. .. James A. Clooney, Syracuse, N. Y.
BOSCOBEL ISLAND—one-half acre, owned by.... .... G. L. Hopkins, Kansas
BLUFF ISLAND—twenty-five acres, owned by.....................E. R. Washburn, New York
CLINTON'S No. 1—fifteen acres, owned by .. .. N. Seely, New York
CLINTON'S No. 2—three acres, owned by.. .... N. Seely, New York
PINE ISLAND—five acres, owned by........ . ...J. B. Hamilton, New York
GOVERNOR'S—three acres, owned by .. .. Hon. T. G. Alvord, Syracuse, N. Y.

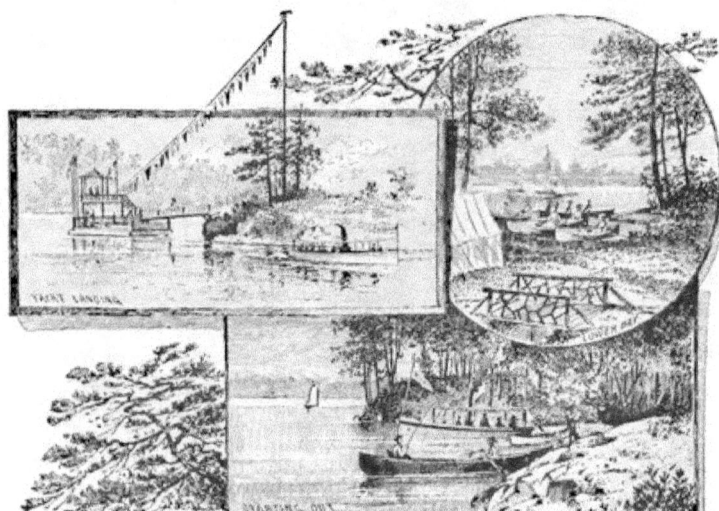

CALUMET—three acres, owned by ....,Chas. G. Emory, New York
LONE ROCK—one acre, owned by........ .. .. W. F. Wilson, Watertown, N. Y.
HEMLOCK—twenty acres, owned by... .. ..Hon. W. F. Potter, W. F. Wilson, Watertown; Hon.
Henry Spicer, Perch River, and others
GUN ISLAND—half acre, owned by............ .... H. H. Warner, Rochester, N. Y.

ETHELRIDGE—(Head of Round Island) owned by ....Dr. Geo. D. Wheeler, Syracuse, N. Y
HAYS COTTAGE—(Head of Round Island) owned by............. .. ..Jacob Hays, New York
VAN WAGENEN COTTAGE—(Head of Round Island) owned by ....H. Van Wagenen, New York
BELDEN COTTAGE—(Head of Round Island) owned by... ....J. J. Belden, Syracuse, N. Y
SHADY LEDGE—(Foot of Round Island) owned by....................Frank H. Taylor, Philadelphia
BROOKLYN HEIGHTS—(Foot of Round Island) owned by...............C. A. Johnson, Brooklyn, N. Y
STEWART, OR JEFFERS—ten acres, owned by.......E. P. Gardiner, Syracuse, N. Y.; John Rogers and
      Miss Haskell, Carthage, N. Y.; L. J. Burdette, Otsego Camp Club; Caleb
      Clark, Cooperstown, N. Y.; Miss E. M. Griswold, Adams, N. Y.; Wesley M.
      Rich, Joseph Sayles, Rome, N. Y.; Reuben Fuller, Chas. Ellis, Clayton, N.
      Y.; H. E. Chickering, Copenhagen, N. Y.; Dr. W. G. Smith, Carthage, N.
      Y.; S. E. Stanton, C. O. Pratt, Syracuse, N. Y.
TWO IN EEL BAY—two acres, owned by....................... ......Dr. E. L. Sargent, Watertown, N. Y
WHORTLEBERRY ISLAND—two acres, owned by.................... ....Mrs. Etta Stillwell, New York
LITTLE WHORTLEBERRY ISLAND—half acre, owned by......Mrs. Lena E. B. Brown, Wilberham, Mass
HUB ISLAND—one acre, owned by........... ....Geo. W. Best, Oswego, N. Y
ONE TREE ISLAND—half acre, owned by. ....Rev. Mattison W. Chase, Gouverneur, N. Y
MAPLE ISLAND—ten acres, owned by.................................Jos. Atwell, Syracuse, N. Y
TWIN—one acre, owned by ....J. L. Huntington, Watertown, N. Y
WATCH—one acre, owned by..........................Mrs. Elizabeth Skinner, New York
ISLE HELENA—one acre, owned by.........................Mrs. Helen S. Taylor, New York
OCCIDENT AND ORIENT—three acres, owned by...................E. W. Washburne, New York
ISLE OF PINES—two acres, owned by ....Mrs. E. N. Robinson, New York
FREDERICK—two acres, owned by. ....C. L. Frederick, Carthage, N. Y
VANDERBILT ISLAND—three acres, owned by...............J. B. Hamilton, New York
BAY SIDE—one acre, owned by....................H. F. Mosher, Watertown, N. Y
LATTIMER ISLAND—one acre, owned by ....Dr. Charles E. Lattimer, New York
RIVER SIDE—(Main Land) one acre, owned by............James C. Lee, Gouverneur, N. Y
KILLEEN'S POINT—(Main Land) one acre, owned by....................—— Killeen, Lockport, N. Y
HOLLOWAY'S POINT—(Main Land) one acre, owned by...... ....Nathan Holloway, Omar, N. Y
FISHER'S LANDING—(Main Land) two acres, owned by.....Mrs. R. Guffee, Miss Newton, Omar, N. Y
ISLAND HOME—one acre, owned by ....Mrs. S. D. Hungerford, Adams, N. Y
HARMONY—one-fourth acre, owned by.....................Mrs. Celia Berger, Syracuse, N. Y
WAVING BRANCHES—(Wells Island) owned by....D. C. Graham, Stone Mills, N. Y.; A. Snell, La-
      fargeville, N. Y.; J. Petrie, Watertown, N. Y.; Jerome B. Loucks, Lafargeville,
      N. Y.; Isaac Mitchell, L. Hughes, Stone Mills, N. Y.; L. Ainsworth, F. Smith,
      H. S. Tolles, Ira Traver, Watertown, N. Y.
BONNIE EYRIE—(Wells Island) owned by.............................Mrs. Peck, Boonville, N. Y
FERN CLIFF—(Wells Island) seven lots, owned by ....Drs. J. S. and C. E. Latimer, New York
GOOSE ISLAND—quarter acre, owned by...... .. ..Mrs. Lottie Simonds, Watertown, N. Y
BAY VIEW—owned by...................... ... .C. S. Lyman, Westmoreland, N. Y
JOLLY OAKES—(Wells Island) two acres, owned by....Prof. A. H. Brown, Dr. N. D. Ferguson, John
      Norton, O. T. Green, Carthage, N. Y.; Hon. W. W. Butterfield, Redwood, N. Y.
PEEL ISLAND—two acres, owned by....Mrs. Sarah P. Lake, Mrs. Jane E. Tomlinson and Miss Maggie
      Parker, Watertown, N. Y.
ISLAND KATE—one acre, owned by.......................Miss Kate Tomlinson, Watertown, N. Y
JOSEPHINE—two acres, owned by ....Mrs. Emma Kenyon, Watertown, N. Y
CALUMET—one-half acre, owned by .... Oliver Green, Boston, Mass

POINT VIVIAN—(Main Land) ten acres, owned by....Rezst Tozer, J. J. Kinney, E. O. Hungerford,
      George Ivers, Evans Mills, N. Y., and others,

LINDER'S—one acre, owned by...John Linder, Utica, N. Y
ISLAND ROYAL—one acre, owned by.................................. Royal E. Deane, New York
CEDAR—one acre, owned by...J. M. Curtis, Cleveland, Ohio
WILD ROSE—one acre, owned by ....Hon. W. G. Rose, Cleveland, Ohio
GYPSY ISLAND—two acres, owned by.............................J. M. Curtis, Cleveland, Ohio
ALLEGHENY POINT—(Main Land) two acres, owned by...................J. S. Laney, Cleveland, Ohio
PHOTO—two acres, owned by. ....A. C. McIntyre, Alexandria Bay, N. Y
BELLE ISLAND—quarter acre, owned by. ....Rev. Walter Ayrault, Geneva, N. Y
SEVEN ISLES—five acres, owned by.................Hon. Bradley Winslow, Watertown, N. Y
LOUISIANA POINT—(Wells Island) three acres, owned by........Hon. D. C. LaBatt, New Orleans, La
QUARTETTE ISLAND—quarter acre, owned by..........................Mrs. Wm. Egan, Chicago, Ill

SUNNYSIDE,

the summer home of Rev. George
Rockwell, now of New York city,
but best known in this region as for
more than twenty years the pastor
of the Reformed Church, the first
organized at Alexandria Bay.

SHADY COVERT—one acre, owned by........................Hon. John C. Covert, Cleveland, Ohio
HILL CREST—(Wells Island) one acre, owned by............ .....Gen. L. H. Shields. Washington, D. C
AVALON—one acre, owned by........................... ..... Mrs. E. D. Beers, Washington, D. C
BEERS—half acre, owned by..............................Mrs. E. D. Beers, Washington, D. C
BELLA VISTA LODGE—(Main Land) five acres, owned by...........Wm. Chisholm, Cleveland, Ohio

NEMAH-BIN—two acres, owned by.............................James H. Oliphant, Brooklyn, N. Y
COMFORT—two acres, owned by.................................A. E. Clark, Chicago, Ill
WARNER ISLAND—one acre, owned by....................H. H. Warner, Rochester, N. Y
MIXIUM—owned by...Rev. W. W. Walsh, Medina, N. Y
LITTLE GEM—owned by......................................Miss Virginia Walton, Alexandria Bay, N. Y
ISLAND GRACIE—owned by.....................................Miss Grace M. Pox, Alexandria Bay, N. Y
WAU WINET—one-half acre, owned by.........................C. E. Hill, Chicago, Ill
CUBA—one acre, owned by...Cornwall Bros
DEVIL'S OVEN—one acre, owned by.............................H. R. Heath, Brooklyn, N. Y
SUNNY SIDE—(Cherry Island) five acres, owned by.............Rev. Geo. H. Rockwell, New York
STUYVESANT LODGE—(Cherry Island) four acres, owned by.........James E. Easton, Brooklyn, N. Y

FISHING PARTIES AT FROST ISLAND.

MELROSE LODGE—(Cherry Island) nine acres, owned by.............A. B. Pullman, Chicago, Ill
INGLESIDE—(Cherry Island) owned by...........................Mrs. G. B. Marsh, Chicago, Ill
SAFE POINT—(Wells Island) four acres, owned by.............H. H. Warner, Rochester, N. Y
CRAIG SIDE—(Wells Island) owned by...........................H. A. Laughlin, Pittsburg, Pa
WEST POINT—(Wells Island) seven acres, owned by.............Henry W. King, Chicago, Ill
PALISADE POINT—(Wells Island) five acres, owned by.............A. C. Beckwith, Utica, N. Y
PULLMAN—three acres, owned by...................................Geo. M. Pullman, Chicago, Ill
NOBBY—three acres, owned by....................................H. R. Heath, Brooklyn, N. Y
LITTLE ANGEL—one-eighth acre, owned by.......................W. A. Angell, Chicago, Ill

WELCOME—three acres, owned by..............................Hon. S. G. Pope, Ogdensburg, N. Y
FRIENDLY—three acres, owned by........................E. W. Dewey, Brooklyn, N. Y
LINLITHGOW—one-fourth acre, owned by ....Hon. R. A. Livingston, New York
FLORENCE—two acres, owned by.................. .. H. S. Chandler, New York
ST. ELMO—three acres, owned by....................Nathaniel W. Hunt, Brooklyn, N. Y
FELSENLCK—(Wells Island) owned by........... ... .Prof. A. G. Hopkins, Clinton, N. Y
POINT LOOKOUT—(Wells Island) one acre, owned by ....Miss L. J. Bullock, Adams, N. Y
EDGEWOOD PARK—(Main Land) thirty acres, owned by.............Edgewood Park Association
EDGEWOOD—(Point Main Land) one acre, owned by................G. C. Martin, Watertown, N. Y
WEST VIEW—(Point Main Land) one acre, owned by.............Hon. S. G. Pope, Ogdensburg, N. Y
VILULA—(Point Main Land) half acre, owned by ....H. Sisson, Watertown, N. Y
NUT SHELL—(Point Main Land) five acres, owned by....Mrs. C. W. Crossmon, Alexandria Bay, N. Y
ISLE IMPERIAL—one acre, owned by.......................Charles I. Singer, Chicago, Ill
HUB CLARK ISLAND—quarter acre, owned by. ... ...Will Clark, Jersey City, N. J
FERN—one acre, owned by............. ....N. and J. Winslow, Watertown, N. Y
HART'S—five acres, owned by..........................Hon. E. K. Hart, Albion, N. Y

DESHLER—fifteen acres, owned by.......................... ....W. G. Deshler, Columbus, Ohio
NETTS—one acre, owned by.......................Wm. B. Hayden, Columbus, Ohio
BONNIE CASTLE—(Point Main Land) fifteen acres, owned by........ ...Mrs. J. G. Holland, New York
CRESCENT COTTAGES—(Main Land) ten acres, owned by...........Bleecker Van Wagenen, New York
POINT MARGUERITE—(Main Land) thirty acres, owned by... ....Mrs. E. Anthony, New York
THE LEDGES—(Main Land) fifteen acres, owned by.......... ....Mrs. Sara E. K. Hudson, New York
LONG BRANCH—(Point Main Land) ten acres, owned by.........Mrs. C. E. Clarke, Watertown, N. Y
SUN-DEW ISLAND—one acre, owned by........ ....Chas. M. Slamm, Paymaster U. S. Navy

MANHATTAN—five acres, owned by.................J. L. Hasbrouck and Hon. J. C. Spencer, New York
ST. JOHN'S—six acres, owned by..Hon. Chas. Donahue, New York
MAPLE—six acres, owned by..J. L. Hasbrouck, New York
FAIRY LAND—twenty acres, owned by.......Chas. H. Hayden and Wm. B. Hayden, Columbus, Ohio
LITTLE FRAUD—one-half acre, owned by...R. Pease, Geneva, N. Y
PIKE ISLAND—one acre, owned by...Frank F. Dickinson, New York
HUGUENOT—two acres, owned by.................................Levi Hasbrouck, Ogdensburg, N. Y
RESORT—three acres, owned by..W. J. Lewis, Pittsburg, Pa
DEER—forty acres, owned by...Hon. S. Miller, New Haven, Conn
ISLAND MARY—two acres, owned by................................Wm. L. Palmer, Carthage, Oak
LOTUS LAND—..
WALTON—seven acres, owned by.................J. N. Robbins and G. H. Robinson, New York
IDLEWILD—four acres, owned by...Mrs. R. H. Eggleston, New York
LITTLE LEHIGH—one acre, owned by..............W. A. and B. H. Wilbur, South Bethlehem, Pa
SPORT—four acres, owned by...E. P. Wilbur, Bethlehem, Pa
SUNNY SIDE—two acres, owned by...W. Stevenson, Sayre, Pa
SUMMER-LAND—ten acres, owned by the "Summer-Land Association," composed of the following
    members: Rev. Asa Saxe, D. D., Francis M. McFarlin, James Sargeant, Emory B.
    Chase, Leon E. Brace, Isaiah F. Force, Henry C. Wisner, Lewis P. Ross, Charles W.
    Gray, George A. Newell, Henry O. Hall, Joseph A. Stul and Frank W. Hawley, of
    Rochester, N. Y.; Rev. Almon Gunnison, D. D., and Frank Sperry, of Brooklyn;
    Rev. Richmond Fisk, Alfred Underhill and Horace Bronson, of Syracuse, N. Y.
ARCADIA AND INA—five acres, owned by...S. A. Briggs, New York
SPUYTEN DUYVEL—one acre, owned by..............................Alice P. Sargent, New York
DOUGLAS—five acres, owned by..Douglas Miller, New Haven, Conn
KIT GRAFTON—one-half acre, owned by.......................Mrs. S. L. George, Watertown, N. Y
LOOKOUT—two acres, owned by...Thomas H. Borden, New York
ELLA—one-fourth acre, owned by..........................R. E. Hungerford, Watertown, N. Y
LITTLE CHARM—one-eighth acre, owned by...............Mrs. F. W. Barker, Alexandria Bay, N. Y
EXCELSIOR GROUP—five acres, owned by..............................C. S. Goodwin, New York
ELEPHANT ROCK—one-eighth acre, owned by.......................T. C. Chittenden, Watertown, N. Y
SUNBEAM GROUP—one acre, owned by...............................Odd Fellows, Watertown, N. Y
ALICE—two acres, owned by...Col. A. J. Casse, New York
SCHOONER—six acres, owned by.................................J. Norman Whitehouse, New York
BIRCH—seven acres, owned by...W. J. Lewis, Pittsburg, Pa
DISGLESPEIL—four acres, owned by...........................Joseph Babcock, Alexandria Bay, N. Y
OCRIS—three acres, owned by.................................Mrs. M. Carter, Poughkeepsie, N. Y
LONG PINE ISLAND—one acre, owned by.......Wm. M. Comstock and Walter Rulison, Evans Mills
HELEN'S ISLAND—owned by.................................Mrs. O. G. Staples, Washington, D. C
ROB ROY—two acres, owned by...A. H. Greenawalt, Pittsburg, Pa
LITTLE DELIGHT—two acres, owned by...............................Louis W. Morrison, New York
CLOUD REST—four acres, owned by..................................A. H. Greenawalt, Pittsburg, Pa
CHILLON ISLAND—four acres, owned by......................A. H. Greenawalt, Pittsburg, Pa
LILY'S ISLAND—quarter acre, owned by.......................Miss L. B. H. Morrison, Erie, Pa
BERKSHIRE—twenty acres, owned by..........................Hon. S. G. Pope, Ogdensburg, N. Y

ON THE ST. LAWRENCE.

By GEORGE C. BRAGDON

AWAY ! away ! the golden day
  Beams brightly on the river,
And time beguiles where happy isles
  Rest peacefully forever;
  And smilingly forever,
  Invitingly forever.

Where isles of green o'erlook the sheen
  Of fair St. Lawrence river,
The silver sheen round isles of green,
  Upon St. Lawrence river.

Ah ! fair the isles, adorned with smiles
  To greet the wooing river;
We float between, 'neath branches green,
  And long to float forever,
  To dream and float forever,
  Forgetfully forever.

With line and boat to dream and float
  On blue St. Lawrence river,
To dream and float with line and boat
  Adown St. Lawrence river.

ON THE REEDS NEAR WELLS ISLAND.

Now dipping oar recedes the shore,
  And on the restless river
We gaily ride, we bound and glide,
  While sunbeams flash and quiver,
  Around us flash and quiver,
  From billows flash and quiver.

And all is bright and care is light
  On old St. Lawrence river;
And care is light and all is bright
  Upon St. Lawrence river.

Shall we forget the friends we met
  And loved upon the river?—
Its songs and dreams and changing gleams?
  No, never; and no, never,
  We shall forget them never,
  We can forget them never.

The thousand joys and sweet alloys,
  Of dear St. Lawrence river,
With sweet alloys the thousand joys
  Of Thousand Island River.

ROUTES TO THE BAY.

Leave New York (Grand Central depot) by N. Y. C. & Hudson R. R. R., through sleeping car over R. W. & O. R. R. via Utica to Clayton, where steamers for Alexandria Bay connect with all trains.

Or you can leave the Central at Utica and take the cars on R. W. & O. R. R., which will take you without change of cars to Clayton, in four and a half hours, 12 miles from Alexandria Bay, where a steamer will be found which will complete the journey in one hour.

Leave the N. Y. Central at Rome, and enter the palace cars of the Rome, Watertown & Ogdensburg railroad. A few hours' ride on these will bring you to Cape Vincent, thirty miles from Alexandria Bay, where steamers run to and fro twice a day, connecting closely with the trains.

Take the West Shore route via Utica in connection with R. W. & O. R. R., or via Syracuse in connection with the Rome, Watertown & Ogdensburg R. R.

Or, if you please, take the other branch at Watertown, and ride through a picturesque country to Ogdensburg, (six hours from Rome,) and there take steamer up the river to the Bay, 36 miles.

Or leave the Central at Syracuse, (which shortens the distance for parties from the west,) and take the Syracuse Northern to Richland, from which place the route is again on the R. W. & O. R. R.

On starting from Oswego (to which city is a railroad from Syracuse and lines of steamers from all the principal points of the great lakes, some of which go to the Bay,) a branch of the R. W. & O. connects with the main road to Richland.

Or if from the east, you take the Delaware & Hudson at Troy or Albany, going through Saratoga and along the west shore of Lake Champlain, to Rouse's Point, there taking the Ogdensburg & Lake Champlain R. R. to Ogdensburg, having a delightful sail from Ogdensburg by steamer to Alexandria Bay.

Or leaving Albany or Troy via D. & H. C.Co., taking steamers through Lakes George and Champlain (the most delightful of all) to Plattsburgh, D. & H. to Rouse's Point, O. & L. C. and steamer to Alexandria Bay, making one of the best trips in this country.

Or from Chicago and the west, take the Limited Express via Chicago and Grand Trunk R. R. at 3.25 p. m. daily, with Pullman sleeper, arriving at Gananoque Junction at 3 p. m. next day, connecting with 1000 Island Railway and steamer for Alexandria Bay, arriving in time for supper. Only 25 hours from Chicago to Alexandria Bay. The " boss route."

Or take Atlantic Express at 8:15 p.m. daily, arriving at Kingston next day at 2 a. m., (except Sundays.) Pullman sleeper runs direct to Kingston wharf and lies over to enable passengers to have a full night's rest, connecting at 5 a. m. with steamers of the Richelieu and Ontario Navigation Co. for Alexandria Bay and Montreal. See map and advertisement.

Or from Portland, Old Orchard Beach, Montreal or Quebec and Maine resorts, take the Grand Trunk R. R. to Brockville, Gananoque or Kingston, and steamers from those points to Alexandria Bay, making one of the most delightful trips in this country.

Or starting from New York, take the New York, Ontario & Western R. R. from West 42d st., Cortlandt or Desbrosses sts, ferries, and enter the through Pullman Buffet sleeping cars for Cape Vincent; (this is the only route from New York running Pullman sleepers to the islands.)

At Cape Vincent the new steamer St. Lawrence makes close connection with the trains, running thirty miles down the river, through the islands, to the Bay.

Connections with the Pennsylvania R. R. by this route are made in Jersey City, in Union station, and all transfer across New York city avoided.

From Portland, Old Orchard Beach and Maine resorts, take the Portland & Ogdensburg R. R., passing through the White Mountains and Vermont, via Rouse's Point to Ogdensburg, and steamer to Alexandria Bay. This is the shortest line from the White Mountains and Maine.

Visitors from the east, whose route is by the Northern Railroad, which connects with the Vermont Central, will take a steamer at Ogdensburg for the rest of the journey, which leaves upon the arrival of train, reaching the Bay in time for supper.

The Royal Mail line of steamers run from Niagara Falls to Montreal, passing down the St. Lawrence by daylight, and stopping at the Bay.

Since the completion of the Lake Ontario Shore Railroad, facilities for reaching Alexandria Bay from the west have improved. Parties may now leave Niagara Falls in palace cars in the morning and ride in them to Cape Vincent, and there taking a steamer, arrive at the Bay in time for supper.

6

HOTELS EN ROUTE.

The following hotels, among others, are recommended to persons en route to the Thousand Islands, on account of their accommodations and management :

BAGG'S HOTEL, Utica, N. Y..T. R. Proctor, Proprietor
GLOBE HOTEL, Syracuse, N. Y.......................................Dickinson & Austin, Proprietors
POWERS HOTEL, Rochester, N. Y..Buck & Sanger, Proprietors
OSBORNE HOUSE, Auburn, N. Y...J. E. Allen, Proprietor
SEYMOUR HOUSE, Ogdensburg, N. Y.......................................F. J. Tallman, Proprietor
DANIELS HOTEL, Prescott, Ont...L. H. Daniels, Proprietor
RUSSELL HOUSE, Ottawa, Ont...James Guin, Proprietor
ST. LAWRENCE HALL, Montreal, Que.,.......................................H. Hogan, Proprietor
FOUTER'S HOTEL, Plattsburg, N. Y.......................................A. J. Sweet, Proprietor
FERGUSON HOUSE, Malone, N. Y..S. E. Flanagan, Proprietor
WINDSOR HOTEL, Montreal, Que...Geo. W. Sweet, Manager
SPRING HOUSE, Richfield Springs...T. R. Proctor, Proprietor

DISTANCE CARD.

Niagara to Toronto,	40 Miles	Montreal to New York	106 Miles
Toronto to Alexandria Bay	167 "	" Albany,	251 "
Oswego to Alexandria Bay	100 "	" Troy,	251 "
Clayton to Alexandria Bay	12 "	" Saratoga	242 "
Alexandria Bay to Montreal,	169 "	" White Mountains,	201 "
" " Watertown	28 "	Ogdensburg to Ottawa,	55 "
" " Utica,	152 "	Montreal to Quebec	280 "
" " Brockville,	21 "	Ogdensburg to Malone,	61 "
" " Portland via O. & L.C.	100 "	" Chateaugay,	53 "
" " Boston via O. & L.C.	442 "	" Chateaugay Chasm,	71 "
" " Ogdensburg	36 "	" Ralph's ,	88 "
Montreal to Portland	278 "	" Saratoga	255 "

TOURISTS' IDEAL ROUTE,

ROME, WATERTOWN & OGDENSBURG RAILROAD.

Great Highway and Favorite Route for

Fashionable Pleasure Travel.

Only All-Rail Route to Thousand Islands.

1889. **New Fast Trains Avoiding Stops.** **1889.**

WAGNER PALACE SLEEPING CARS.

NEW YORK AND PAUL SMITH'S, 15 HOURS.

NEW YORK AND CLAYTON, 11 HOURS.

NIAGARA FALLS AND CLAYTON, 9 HOURS.

NIAGARA FALL, SPORTLAND and BAR HARBOR, Me.,

Via Norwood, Fabyans, Crawford Notch and all White Mountain Resorts.

WAGNER PALACE DRAWING - ROOM Cars.

NIAGARA FALLS AND CLAYTON.

ROCHESTER AND CLAYTON.

SYRACUSE AND CLAYTON.

ALBANY AND CLAYTON.

UTICA AND CLAYTON.

Direct and immediate connections are made at Clayton with powerful steamers for Alexandria Bay and all Thousand Island Resorts; also with Rich and Ont. Nav. Co.'s steamers for Montreal, Quebec and River Saguenay, passing all of the Thousand Islands and Rapids of the River St. Lawrence by daylight. For tickets, time tables and further information, apply to nearest ticket agent or correspond with General Passenger Agent, Oswego, N. Y.

ROUTES AND RATES FOR SUMMER TOURS.

A beautiful book of 200 pages, profusely illustrated, contains maps, cost of tours, list of hotels, and describes over 100 Combination Summer Tours via Thousand Islands and Rapids of the St. Lawrence River, Saguenay River, Gulf of St. Lawrence, Lake Champlain, Lake George, White Mountains, to Portland, Kennebunk, Boston, New York and all Mountain, Lake, River and Sea Shore Resorts in Canada, New York and New England. It is the best book given away. Send ten cents postage to General Passenger Agent, Oswego, N. Y., for a copy before deciding upon your summer trip.

E. S. BOWEN,
Acting General Manager.

THEO. BUTTERFIELD,
General Passenger Agent, OSWEGO, N. Y.

The CENTRAL VERMONT RAILROAD

IS THE FASHIONABLE AND SCENIC ROUTE FOR

TOURIST and PLEASURE TRAVEL

BETWEEN

MONTREAL, BOSTON, NEW YORK,

AND ALL SUMMER RESORTS IN NEW ENGLAND AND CANADA, INCLUDING

The GREEN, WHITE and ADIRONDACK MOUNTAINS,

Lakes Champlain & George, Bar Harbor, Me., Block Island & Newport, R. I.

DOUBLE DAILY FAST EXPRESS TRAIN SERVICE,

With Pullman and Wagner Palace Buffet Parlor and Sleeping Cars between

Montreal & Boston, Montreal and New York, Montreal and the White Mountains

Norwood, N. Y., and Bar Harbor, Me.,

Through the WHITE MOUNTAINS BY DAYLIGHT, are advantages only offered by this popular and old established route through the

SWITZERLAND OF AMERICA.

———

CORNWALL BROTHERS, Ticket Agents, Alexandria Bay, N. Y.

A. C. STONEGRAVE, Canadian Passenger Agent, 131 St. James-St., Montreal, Que.

J. W. HOBART, St. Albans, Vt.
General Manager.

S. W. CUMMINGS,
Gen'l Passenger Agent.

50   MILES   50

AMONG THE ISLANDS

More Islands to be Seen,

MORE MILES TO RIDE,

And more Pleasure to Enjoy by taking a Trip on the

NEW ✦ ISLAND ✦ WANDERER,

**Just completed, and built expressly for this Route, than
can be had on any other boat on the river.**

Don't Miss this Trip.   ☞FARE 50 CENTS.

CAPT. E. W. VISGER, Manager.

THE - NEW - DIRECT - LINE

BETWEEN

ALEXANDRIA BAY,

THOUSAND ISLANDS.

—AND—

THE ADIRONDACK WILDERNESS,

IS VIA THE

CENTRAL VERMONT & NORTHERN ADIRONDACK

RAILROADS.

THE ALL RAIL LINE TO THE HEART OF THE GREAT NORTHERN WILDERNESS.

VIEWS OF CHATEAUGAY CHASM SCENERY ON

SPRUCE PASS—RAINBOW FALLS.

GIANT GORGE—PULPIT ROCK.

Ogdensburg and Lake Champlain Division Central Vermont Railroad.

CENTENNIAL HALL,

ALEXANDRIA BAY, N. Y.

One of the most attractive features at Alexandria Bay is Centennial Hall. It is a magnificent structure, in the style of a Swiss cottage, 60x14 feet in size, entirely surrounded by a broad veranda 8½ feet wide, making the entire dimensions 77x31 feet, thus affording a delightful, uninterrupted promenade of 216 feet.

The entire finishing and furnishing is of the richest description. Its sides are made up of windows, from each of which is a fine view. At each end are windows of stained glass. Flagstaffs surmount the edifice, bearing the respective banners of the United States and England. Well, you ask, what is all this for? Just what we are coming at. Here will be kept

ALL THE DELICACIES OF THE SEASON.

Here you will find the most delicious of ice-creams, made of *cream*, too, my dear madam. Think of an iced lemonade in this delightful spot! Perhaps it is some of those fresh, tempting oranges, pineapples, peaches or bananas that you prefer. If it be anything in the line of fruits, or the most tempting of confectionery, they are here. Here, too, is the

CHOICEST LITERATURE OF THE DAY.

Books, papers, magazines, etc., and McIntyre's Gems of the Thousands Isles are had here, and, in fact, much of all that goes to make life pleasant as well as profitable. In a word, Centennial Hall is *un Grande Place du Resort.* Do not fail to visit it.

GILSEY ✦ HOUSE,

COR. BROADWAY AND 29th STREET,

NEW YORK CITY.

ON THE

EUROPEAN PLAN

LOCATION CENTRAL.
CUISINE & APPOINTMENTS UNSURPASSED.

J. H. BRESLIN & BRO.,

PROPRIETORS.

Missouri Pacific Railway.

THE FAST MAIL ROUTE.

St. Louis to Kansas City, Pueblo, Denver, Salt Lake City, San Francisco.

IRON MOUNTAIN ROUTE FOR ALL POINTS IN

Arkansas, Texas, Mexico and California.

THE ONLY DIRECT ROUTE TO THE FAMOUS HOT SPRINGS, ARKANSAS.

Central Avenue, Hot Springs, Arkansas.- View from North Mountain, photo. by Kennedy.

Pullman Buffet Sleeping Cars and Free Reclining Chair Cars

ON ALL TRAINS.

For Rates of Fare, Time Tables, Descriptive Books of Hot Springs, and any further information, write

H. C. TOWNSEND,
Gen'l Passenger and Ticket Agent.
ST. LOUIS, Mo.

WM. E. HOYT,
Gen'l Eastern Passenger Agent,
391 BROADWAY, NEW YORK.

THE THOUSAND ISLAND

——AND——

ST. LAWRENCE RIVER STEAMBOAT COMPANY,
(LIMITED,)

IN CONNECTION WITH THE

Rome, Watertown & Ogdensburg
RAILROAD.

Only Direct Route Between

Cape Vincent, Clayton, Alexandria Bay, Gananoque and Kingston.

STEAMERS:

St. Lawrence, Islander, Maynard, Princess Louise, Maud and Pierrepont.

8

NEW YORK, ONTARIO AND WESTERN RAILWAY CO.

ONTARIO ROUTE.

SEASON ——————— OF —————— 1889.

ONLY LINE RUNNING THROUGH

PULLMAN PALACE BUFFET SLEEPING CARS

BETWEEN NEW YORK AND THE THOUSAND ISLANDS.

The Pullman Buffet Sleepers run on this line are of the latest model, and are the most magnificent cars put in the public service.

DEPOTS AND FERRIES IN NEW YORK

At foot of WEST 42d STREET and JAY STREET.

Thousand Island Express leaves New York, 42d-st. depot, at 5:15 P. M., Jay-st. 5:15 P. M., arriving at Cape Vincent 10:30 A. M. and at Alexandria Bay, via Steamer St. Lawrence, at 12:30 P. M.—running twenty-five miles down the river through the entire length of the Thousand Islands.

New York Express leaves Alexandria Bay, via Steamer St. Lawrence, at 11:10 P. M., leaving Cape Vincent at 1:01 P. M.; arrives at New York at 9:25 A. M.

Through PULLMAN SLEEPING CARS between Cape Vincent and New York.

All trains via the "Ontario Route" run along the picturesque West Shore of the Hudson, through the Highlands, over the foothills of the Catskills, and through the mountain regions of Central New York, as well as through the beautiful valleys of the Delaware, Susquehanna and Chenango Rivers, making the landscape route across the Empire State.

TOURISTS' TICKETS ON SALE AT ALL OFFICES,

Embracing Trips to Niagara Falls, Lake Regions of Canada, Thousand Islands, Montreal, Quebec, Lake Champlain, White Mountains, Etc., Etc.

Time Tables, Tickets and Information Furnished at any of the Company's Offices Below:

In Brooklyn—No. 4 Court street, No. 808 Fulton street, No. 829 Fulton street, Brooklyn Annex office, foot of Fulton street; in Broadway, Williamsburg: 220 Manhattan avenue, Greenpoint. In New York city—No. 364 Broadway, corner Franklin street; No. 916 Broadway, near Madison Square; No. 747 Sixth avenue, corner of Forty-second street; No. 1,323 Broadway, near Thirty-third street; No. 116 East 125th street, Harlem; No. 56 Broadway, World Travel Company; N. Y., O. and W. Railway, foot of Forty-second street.

Agents of the New York Transfer Company, New York, will furnish tickets, and check baggage from residence to destination.

Send for a copy of "Summer Homes" along the New York, Ontario and Western Railway, with full list of Summer Hotels, Boarding Houses, terms, etc. This book is replete with valuable information, and is furnished free on application.

J. E. CHILDS, Gen'l Manager.   J. C. ANDERSON, Gen'l Passenger Ag't.

Post Building, 16 and 18 Exchange Place, New York.

❋PEOPLE'S EVENING LINE STEAMERS.❋

DREW AND DEAN RICHMOND.

CAPT. S. J. ROE.   CAPT. THOS. POST.

ELECTRIC  IN EVERY

LIGHTS   ROOM.

Leave ALBANY for NEW YORK Every Week Day at 8 P. M., or on arrival of trains from the North, East and West.

TICKETS SOLD AT STATIONS OF THE ROME, WATERTOWN & OGDENSBURG R. R., N. Y. C. & H. R. R. R., AND WEST SHORE R. R., VIA THE PEOPLE'S LINE, FOR ALL POINTS SOUTH, AND BAGGAGE CHECKED THROUGH.

Leave NEW YORK for ALBANY Every Week Day from Pier 41 North River, foot of Canal Street, at 6 P. M., arriving at Albany next A. M., connecting with trains of the New York Central R. R., Rome, Watertown & Ogdensburg R. R., for the West and Thousand Islands, D. & H. C. Co.'s roads for Saratoga, Lake George, Lake Champlain and Adirondacks ; also Howe's Cave, Sharon Springs, and Cooperstown.

J. H. ALLAIRE.   M. B. WATERS,
General Ticket Agent,   General Passenger Agent,
NEW YORK.   ALBANY.